THE GETTY
CONSERVATION
INSTITUTE

November 2007

The Nature of Conservation: A Race Against Time was originally
published by the Getty Conservation Institute (GCI) in collaboration
with the International Council of Museums (ICOM) in 1986. Meant to
serve as an introduction to the field of conservation of cultural
heritage for a general audience, the book was printed in English,
French, and Spanish, and produced in conjunction with the 14th
General Conference of ICOM, held in Buenos Aires, Argentina, in that
year.

This modest volume has been out of print for many years. It has
served for many people as an entrée into the challenges and rewards
of the field of conservation, and much of it is still relevant to the
field. In a relatively young field like conservation, however, 20 years
represents a significant time span, and there are some aspects of the
book that do not reflect current thinking on the topic and the field.
Much good (and certainly no harm) will come of studying The Nature
of Conservation, but the reader should be forewarned that it
represents a moment in time, and that the approach to conservation
has changed in many respects since its publication.

This special re-printing of The Nature of Conservation is at the
request of the Institute of Museum and Library Services (IMLS), who
have chosen it to form part of the core collection of the Connecting
to Collections Bookshelf. We are honored to be part of this
important and inclusive initiative, and believe that you will find this
publication a stimulating introduction to the field of conservation.

Timothy P. Whalen
Director

1200 Getty Center Drive, Suite 700, Los Angeles, California 90049-1684
Phone 310 440.7325 Fax 310 440.7702

The Getty Conservation Institute

The Getty Conservation Institute (GCI), an operating organization of the J. Paul Getty Trust, was created in 1982 to enhance the quality of conservation practice in the world today. Based on the belief that the best approach to conservation is interdisciplinary, the Institute brings together the knowledge of conservators, scientists, and art historians. Through a combination of in-house activities and collaborative ventures with other organizations, the Institute plays a catalytic role that contributes substantially to the conservation of our cultural heritage. The Institute aims to further scientific research, to increase conservation training opportunities, and to strengthen communication among specialists.

Other organizations of the Trust are the J. Paul Getty Museum, the Getty Center for the History of Art and the Humanities, the Getty Art History Information Program, the Getty Center for Education in the Arts, the Museum Management Institute, the Getty Grant Program, and the Program for Art on Film, a joint venture with The Metropolitan Museum of Art.

The Nature of Conservation

BLUE GLASS
LEAD KOHL
COLOURED
GLASS M.M?

"HASTINGS"
CLEAR CASTING
RESIN
(POLYESTER)

"TAYLON" BM4.CO PA
(POLYESTER) EMBEDDING
RESIN + TRANSLUCENT

The Nature of Conservation
A Race Against Time

Philip Ward

The Getty Conservation Institute
Marina del Rey, California

Glass conservation.

The Getty Conservation Institute
initiated the publication of this
book, recognizing the need for
a public forum on conservation
issues. The purpose of this book
is to stimulate dialogue within
the profession and, at the same
time, to acquaint the public with
the philosophies and efforts
of those seeking to preserve the
cultural heritage.

This publication was presented
at the 14th General Conference
of the International Council of
Museums (ICOM) held in Buenos
Aires, Argentina in October 1986.
It was inspired by and a comple-
ment to an exhibition prepared
by the ICOM Conservation
Committee for this conference.

The author, Mr. Philip Ward,
Director of the Information and
Extension Services Division,
Canadian Conservation Institute,
is a distinguished scholar; his
participation in this project was
made possible by the cooperation
of the Canadian Conservation
Institute.

Second printing, 1989

Printed in the United States
of America.

The Getty Conservation
Institute
4503 Glencoe Avenue
Marina del Rey, California
90292-6537

Library of Congress Catalog
Card Number: 86-82718

ISBN 0-941103-00-5

Contents

Preface

This is an outspoken book. The views it expresses are the author's and represent neither those of the many people who assisted him, nor the policies of their organizations. That it was written by a notoriously opinionated elderly conservator may be an explanation for this, but it is not the reason. In addition to a certain cynicism, more than thirty years in museums have imparted some strong convictions. They are: that museums are one of the vital spiritual cornerstones of civilized society; that preservation is the most important thing that they do; and that conservators are at once the most valuable, the most fortunate, and the most frustrating people who work in them.

Conservators are valuable because they possess the skills to preserve the collections that are the museum; they are fortunate because they enjoy the most intimate contact with the objects on which they work, and thus with those who made them; and they are frustrating because they have a tendency to immerse themselves in the technical aspects of their work, virtually to the exclusion of all else.

The technology of conservation is absorbing, but too exclusive a devotion to it has led conservators to adopt the conventions of science in their communications, which many other museologists find incomprehensible. The result is that conservators are becoming increasingly isolated from their colleagues, when their mission, in fact, demands the most fluent communication.

That is why the invitation to write this small book was so irresistible. It offered an opportunity to do what, with a few notable exceptions, we have so far failed to do: to explain ourselves to our colleagues. For the errors and omissions the book undoubtedly contains, I apologize and offer only the excuse that, unavoidably, it was written in great haste. For its bluntness, I offer no apology: That is deliberate. If it generates debate between conservators and other museologists, it will have succeeded, because debate is communication.

Finally, I should like to express my personal gratitude to Mr. Steen Bjarnhof, Head of Department of the Conservation School of the Royal Danish Academy of Art, under whose direction the poster exhibit was assembled. At very short notice, and at the busiest time of the academic year, Mr. Bjarnhof not only gave me access to the poster material and provided working facilities, but he also gave generously of his time, his wisdom, and his friendship.

Two members of Mr. Bjarnhof's staff also deserve special mention. Ms. Berit Møller, who was responsible for the assembly of the poster exhibit, tolerated my intrusion with good humor and, through her fine organization of the material, assisted me greatly. Mr. Mikkel Scharff, similarly, performed many small acts of kindness, which enabled me to complete the work in Copenhagen within the very limited time available.

Philip Ward

Introduction

Photography in the Scientific Examination of Works of Art

Photography provides the conservator with an invaluable record of the features of a work of art that may be transient or invisible to the unaided eye. Scientific examination techniques may be applied to this information at any time in the future.

In addition to an overall "before treatment" view by "flat" light, photographs by raking light, ultraviolet, infrared, and x rays all provide essential working information for the conservator, which may be enhanced by the visible-light microscopy of paint sections.

No one method of examination can reveal all the relevant characteristics of a work of art, but in combination, the techniques now available can reveal features previously known only to the artist himself. Not only can we determine the structure of the painting, the form and extent of previous restorations and the causes of deterioration, but we can also trace the changes made by the painter as he worked.

Unknown artist. De Danske Kongers Kronologiske Samlinger, Rosenborg Slot Copenhagen, Denmark.

This book is not for conservators, but for other museum professionals who seek a perspective on conservation.[1] It is broad, informal, and deliberately unspecific. Above all, it tries to explain the place and function of conservation in the museum. In some institutions, conservation has acquired a mysterious or even, in the perception of some professionals, a threatening aspect when, in fact, it is simply common sense. Such an unfortunate disjunction may be ascribed, not only to the defensiveness that sometimes afflicts the newcomer in an unfamiliar environment, but also to the conservator's preoccupation with his own exacting discipline and the technical vocabulary it uses. The pages that follow are not technical.

By definition, museums have four classic functions: they *collect*, they *preserve*, they conduct *research*, and they *present* or interpret their collections to the public in light of their research. *Preservation* is the most fundamental of these responsibilities, since without it, research and presentation are impossible and collection is pointless. *Conservation* is the technology by which preservation is achieved.

Conservation is the youngest of museum disciplines. Although long used in reference to the preservation of natural resources, the word has only been applied to the preservation of works of art since 1930. In the following year, the first conference on the application of scientific methods to the examination and conservation of works of art took place in Rome under the auspices of the International Museums Office of the League of Nations, the forerunner of the International Council of Museums.

J. Courtois, *Battle Scene*. Before treatment, the deterioration of varnish due to humidity obscures the image. Insert shows detail after removal of damaged varnish.

Visual Restoration of Paintings

There are two primary reasons for restoring a painting: It may have become structurally unsafe due to mechanical damage, i.e., the separation of the paint layer from its underlying support; or it may have become illegible because the image has been obscured.

The latter can occur in several ways. A faulty painting technique may have resulted in a surface so irregular that shadows and reflections mask the image; the surface may be covered by dirt or the varnish may have darkened or bloomed; or some of the paint may have been removed, perhaps by excessive cleaning, leaving the image incomplete. All are classic vices, generally treated by traditional methods.

For at least a century before 1930, many museums employed restorers as staff members or as private contractors. Having neither the knowledge nor the technology to avoid or control deterioration, they naturally pursued a philosophy quite different from that of modern conservators. Objects were used or stored, and when they were damaged, those thought to warrant the expense were restored, while the rest were simply hidden away.

During the 1930s and 1940s, a few institutions in western Europe and the United States began to study the causes of deterioration, to apply their findings to the care of their collections, and to share this information with their restorers. A new philosophy—a philosophy that recognized the need for prevention before repair—began to emerge.

The first General Conference of the International Council of Museums (ICOM), held in Paris in 1948, created the first ICOM Commission (which was on the care of paintings) and brought together directors of major European museums, restorers, and scientists. Subsequently, an ICOM Committee for Museum Laboratories was established, which in 1963, became the ICOM Committee for Conservation.

Meanwhile, in 1950, the International Institute for the Conservation of Historic and Artistic Works (IIC) was founded in London; eight years later it published *The Conservation of Antiquities and Works of Art* by H.J. Plenderleith. The importance of this volume lay in its diagnostic content: It was the first systematic explanation of the mechanisms of deterioration, which remains the fundamental core of conservation knowledge. It also demonstrated the practical possibility of prevention and, by contributing a knowledge of materials science to the skills and traditional wisdom of restoration, it shaped the new discipline of conservation.

Impelled by the explosion of scientific knowledge and an unprecedented worldwide interest in heritage, the growth of the conservation field has been dramatic. With it, the change to a "preservation philosophy" has ensured that museologists no longer accept deterioration as inevitable, but rather seek to protect their collections against damage from any cause.

Nevertheless, restoration remains a vital activity for the conservator. Museums still collect objects that require restoration and, despite our efforts to protect them, objects still deteriorate in museum use. Thus, preservation and restoration are simultaneous museum activities.

Despite the advances that scientific research has brought to restoration, it remains an art. A speaker at the Hungarian Institute of Conservation and Methodology of Museums in 1976 described it in these words:

Restoration is a race against time for the maximum extension of the life of the material, and thus that of the work of art. This determination also implies that restoration is not a branch of science. Its aim is not to attain certain historical or scientific results, but to utilize such results in the interest of the object.

While the impetus for conservation development has always come from the major public museums, restoration continues to be a significant private sector industry. In countries where private collectors can support sufficient numbers of private restorers, museums have sometimes preferred to hire these professionals on contract, in place of having conservators on staff. With such an arrangement, the restorer may not be called upon until the condition of a popular object becomes an embarrassment. In other words, restoration philosophy still survives in these museums as it does, to some extent, in institutions that rely on national or regional conservation services. Since conservators can only perform their preventive function effectively when they have daily contact with the collections, external conservation services and conservators in private practice are effectively limited to restoration.

Obviously, many institutions have little choice in this matter. Small museums with only one or two paid staff could never justify the expense of a conservation facility; but it is difficult to understand the ethical justification for such economy by larger institutions. If a museum accepts its role as a custodian of heritage collections, it must surely accept also the responsibility for their care. It is a matter of priorities.

Yet still, as Jean des Gagniers wrote, in some museums

c'est comme si l'on avait cru que l'important était le geste de "sauver" une collection en l'acquérant; que, cela fait, on pouvait la reléguer n'importe où. Bref, après avoir fait un beau geste, on trouvait la collection vraiment encombrante.

It is as if one had believed that the important thing was the act of "saving" a collection by acquiring it; that, once this was accomplished, one could ignore it. In short, after having made a noble gesture, one found the collection truly a nuisance.
—Jean des Gagniers, *La Conservation du Patrimoine Muséologique du Québec* (Québec, 1982), p. 11.

Crucifix from Edestrup Church, Denmark. Nationalmuseet, Brede, Denmark.

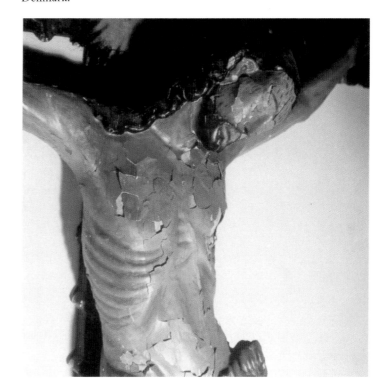

Polychromed Sculpture

In the conservation of poly-chromed sculpture, the problems of preserving and restoring early paintings—often comprising complex techniques and materials—are compounded by their three-dimensional supports, which are frequently made of wood, often large and structurally intricate, and almost always fragile. Furthermore, like church frescoes, their environments may have changed drastically. Perhaps for the first time in a thousand years, some churches are now being heated (intermittently). As a result, supports expand and contract and the adhesion of paint layers is increasingly threatened. As deterioration becomes more obvious, restoration is required. The removal of dirt accumulated during centuries of neglect frequently reveals works of great beauty.

The chapter titled "The Nature of Conservation," attempts, by means of a summary, to define conservation to the non-conservator.

"The Primary Activities of the Conservator" focuses on specific tasks and philosophies as they are applied to the parallel activities of preservation and restoration.

The impact of scientific research upon the discipline of conservation is immeasurable. "The Role of Science in Conservation" describes the application of science in conservation as well as some of the recent technologies that are available to conservators.

The education of conservators largely determines their initial attitudes. "The Training of Conservators" presents a brief history of conservation training and an overview of present-day practices: their strengths and their omissions.

"Conservation Services" examines the ways in which conservation units may be incorporated into the museum organization and discusses some of the difficulties that may be encountered. It goes on to describe the characteristics of the systems by which regional and national conservation services may be delivered.

"The Conservator as Museologist" focuses on interaction with other museum professionals. Benefits and potential problems are frankly discussed and difficulties are analyzed.

"The Future" introduces some of the developments that may be encountered by future conservators.

Salt shaker damaged by excessive humidity. School of Conservation, Royal Danish Academy of Fine Arts, Copenhagen, Denmark.

that the most elaborate or extensive treatments must always be applied. Different *degrees* of treatment may be justified by circumstances, but whatever the extent of treatment, its *quality* must be the best.

- All treatment must be preceded by a thorough technical examination of the object and must not proceed until the conservator is satisfied that all significant information relevant to the proposed action is in his possession.

- Treatments must be recorded completely and honestly, in a form that will be permanently available (as far as it is within the conservator's power to ensure) to future researchers and conservators.

- The preservation of the fabric of the object is not necessarily limited to its original material since earlier repairs or modifications during use may be of great historical significance. The importance of any such feature should be determined jointly with the curator, after careful examination and research.

- New material should be added to the object to the least possible extent and must be compatible with its future welfare.

Pyramid of the Sun, Teotihuacan, Mexico.

- All intervention must respect the integrity of the object. This is one of the most difficult criteria to uphold because it is inherently subjective.

- To the best of his ability the conservator must seek always to maintain the currency of his technical knowledge.

- The conservator must be aware of his own limitations. He should not make determinations that are beyond his knowledge or undertake tasks that exceed his skills.

- General advice on the care of collections is expected from the conservator, but he should offer advice on treatment only to another conservator.

- In treating an object, the conservator continuously performs the following: *examination/recording/diagnosis/action/recording/care.*

2

The Primary Activities of the Conservator

The principles of conservation practice described in the previous chapter represent a professional unanimity that is extraordinary in a discipline so young, so widespread, and so given to forthright debate. They are the practical expression of the fundamental concept that guides all conservation: respect for the integrity of the object. There, however, the unanimity ends. The application of conservation must take differing forms when adapted to the needs of different types of collections and to the curatorial disciplines that administer them. In this chapter, those varying applications are discussed under the broad headings of "Preventive Conservation" and "Restoration": the two primary activities of conservators.

Preventive conservation requires control of the environment. Air, bacteria, and fungi samples taken from Nefertari's tomb, West Thebes, Egypt being packaged for transport.

Preventive Conservation

Preventive conservation is possible only because scientific research has given us a better understanding of some of the mechanisms of deterioration. Although that body of knowledge continues to grow rapidly in detail, the fundamental principles are well established. Deterioration is *not* inevitable and "ageing" is only a multiplier of known and generally controlable causes. The major causes are environmental: light, temperature, humidity, and atmospheric gases. To these may be added mechanical damage due to mishandling and inadequate support; chemical damage due to contact with reactive materials; and biological damage by microorganisms, plants, insects, and animals. All these factors can be controlled, though because some (such as light and air) can rarely be eliminated, deterioration can be greatly retarded, but not completely arrested. Thus, the methodology of preventive conservation is indirect: Deterioration is reduced by controlling its causes.

Advice on environmental standards is often misunderstood because others do not realize that, far from being the arbitrary choice of the conservator, standards are actually determined by the physical properties of materials. Therefore, there is no point in relaxing standards that may be difficult to achieve. In fact, the conservator only states the internationally agreed upon predictions for the response of given materials to specific conditions. To ask him to change his prediction is rather like asking a meteorologist to change his forecast: It won't change the weather.

Manuscript exhibition. The J. Paul Getty Museum.

Lighting and Climate Control

All visible light is damaging and its effect is cumulative. Objects should be stored in the dark, but exposure is unavoidable during periods of exhibition. Nonetheless, damage can be reduced by controlling both the intensity and the duration of exposure to light.

Relative humidity (RH) is a measure of the moisture content of the air, related to the temperature at a given time. While the optimum RH varies for different materials, a compromise suitable for the majority of museum objects is proposed. Extremes should always be avoided, but even within the recommended range, fluctuations should be minimized.

Every activity that affects the physical welfare of museum objects concerns the conservator. Other museum staff may be more directly involved in the various uses of the collections, but they are not trained in the preservation of those collections. The intervention of the conservator in all collections-using activities is not always welcome, especially in museums with a short conservation tradition.

One solution to the problem of shared responsibilities is a management device known as the "troika."[2] It recognizes three distinct areas of responsibility for the use of the collections. The curator is responsible for the intellectual aspects of objects; the conservator for the physical aspects; and the designer, researcher, or educator for the activities in which they are used. If the three participants then jointly and equally contribute to all decisions regarding the use of the collections, satisfactory compromise agreements may be reached.

In assisting the activities of other staff, conservators may also design and sometimes build special supports for fragile objects on exhibition or in storage, or special containers for shipping. These activities, along with monitoring environmental conditions and controlling insect pests, are all noninterventional forms of preventive conservation. Additionally, conservators are expected to disseminate information through publications that reach a broad audience, as well as to educate museum staff in the basics of caring for collections. In these ways, many larger museums share their expertise with the public, with private and corporate collectors, and with museums that have no conservator.

Preventive conservation also includes direct intervention through treatments designed to stabilize deteriorating objects, to consolidate or strengthen weak ones, and to protect vulnerable ones without in any way restoring them. These applications vary with the priorities of each discipline. For example, with works of art the main priority is generally to preserve the image more than the material of which it is made. This may justify the use of protective coatings, such as varnishes, to a greater extent

Structural Restoration of Paintings on Canvas

Structural restoration of a painting may be required when either the paint layer or its support have suffered substantial damage. This damage may take the form of obvious paint losses, tears, or sagging canvas; or it may be due to the deformation of the support or to a loss of adhesion between the paint and the support, causing "cupping" and incipient paint loss.

These problems may result from mechanical damage or material defects, but most often they are the result of deterioration caused by excessive heat or fluctuating humidity. Treatment may require the painting to be flattened on a vacuum table; sometimes it must be relined and the losses must be filled and inpainted.

Damage to photographic negative from improper storage. School of Conservation, Royal Danish Academy of Fine Arts, Copenhagen, Denmark.

Photographic Documents

Most people recognize that photographic negatives are extremely fragile: glass negatives are easily broken, and all may be easily scratched and marked by careless handling. Their suscepti-bility to chemical damage, however, is less well known. Many archives hold thousands of negatives and often receive very large collections that they are not able to examine in detail and are even less able to store correctly. Consequently, negative collections often remain in unsuit-able storage containers for many years.

Water may strip photographic emulsion from glass negatives; airborne pollutants may bleach exposed emulsion; acids in paper envelopes and plasticizers in synthetic materials may attack them chemically; the glues of paper envelopes may bleach them; and notations in ink may "print-through" from the envelope to the negative. It is essential that nega-tives be stored in containers of neutral pH. Glueless envelopes of high-quality pure rag paper with-out buffers or other additives are suitable.

than would be acceptable for other artifacts. Even so, only stable, removable materials are used and their application is carefully recorded. Similarly, since the priority for archives is the preservation of information, archival records may be preserved by copying them in a different medium or onto more durable material. Excavated objects, on the other hand, may contain within their materials or even in the dirt or patina on their surfaces, priceless evidence for research. It may be unacceptable, not only to add any foreign material, but even to clean them, or in some cases, to handle them or to allow them to come into contact with organic material.

Thus, the fine art conservator may varnish a painting to protect it, the archival conservator may photocopy or microfilm a document, and the archaeological conservator may design and build a special noncontaminating container for an artifact. Conservation is sometimes criticized for excessive specialization, and there is no doubt that smaller museums need generalists, but the specializations of conservation only reflect those long established by the disciplines they serve.

Restoration

The purpose of restoration is to repair damage that has already occurred. Damage is irreversible; it can be concealed, broken parts can be rejoined, missing parts can be replaced, and weak parts can be strengthened, but at best, the object will only *appear* to be as it was. The restored object will inevitably be less complete, less original, less honest.

Restorations—indeed all treatments—are cumulative. The preservation of objects in perpetuity implies that they will undergo treatments in the future, as many already have in the past. Thus, there is an ethical imperative for minimizing treatment, since each subsequent intrusion moves the object farther from its original state.

Yet there is no alternative. Restoration may be essential to prevent further deterioration, or it may be necessary in order to make an object usable again as, for example, in the case of a painting that has become obscured. We must accept that the loss of originality that restoration implies is a trade-off against the possible loss of the object itself or its usefulness. Nevertheless, that acceptance carries with it the obligation to keep the loss of originality to a minimum and to record treatment in detail so those who study the object in the future will not be deceived by our work.

Although these rules apply to all treatments, they are of particular concern in restoration because it is a deliberate, considered alteration of the object. In this respect, restoration has a rather special position in conservation: It is the activity that most depends upon the skill, judgment, and sensitivity of the individual, and it also presents the most exacting ethical challenge.

It is difficult to generalize about restoration because, as with preventive treatment, the acceptable degree of intervention varies from one discipline to another. There can be only one *standard* in conservation (the best that circumstances allow), but the *degree* of treatment—including restoration—may depend on such variables as available resources, the future use of the object, and the needs of the particular discipline to which it belongs.

In the repair of books and documents, for example, it is usual to restore bindings and containers, but not to replace missing text.[3] Because the purpose of archival collection is to preserve information, restoration is usually confined to rare documents of intrinsic value, while the contents of others are reproduced in a more durable medium.

Graphic Documents

The conservation of documents on paper is one of the most urgent needs of our time. In many cases, the technology of the recent past survives only in the form of disintegrating technical drawings on the most transient of materials. Tracing papers present special difficulties since they become stiff, discolored, and extremely brittle. Each drawing must be relaxed, cleaned, and lined onto a new support, and often, many detached pieces must be carefully fitted into place.

Older documents lose substantial areas of paper, which must be restored if the document is to be rebound or stored for future reference. The relatively new technique of leaf-casting enables the conservator to duplicate exactly the characteristics of any paper and to replace missing portions without the use of adhesives. Leaf-casting is not only faster and more precise than previous methods, it is also reproducible and self-masking, so that even the most minute lacunae are automatically filled.

Conservation and restoration of a Japanese scroll. School of Conservation, Royal Danish Academy of Fine Arts, Copenhagen, Denmark.

Arabian oryx habitat diorama. Natural History Museum of Los Angeles County.

Natural History Collections

The urgency of the need for the conservation of natural history collections has only recently been appreciated. Formerly, many specimens were repaired by taxidermists or simply replaced when they deteriorated. Today, many of those specimens are irreplaceable and must be preserved.

Geological specimens deteriorate in response to their environment just as the same minerals do when they occur in a "worked" form in artifacts. Pyrite disease damages bones; zoological specimens improperly cleaned and stored are stained by oils; mounted birds, mammals, and insects are attacked by other insects; and specimens stored in alcohol or formaldehyde often lose form and color. An immense new field of hitherto neglected problems confronts the natural history conservator.

The importance of a work of art usually lies in the image that the artist created, rather than in its material. Therefore, while the original material must be preserved to the greatest extent possible, the restoration of the image takes precedence, and the introduction of new materials to that end is perfectly acceptable. Indeed, suitable modern materials are preferred because they are more readily distinguishable and can be removed if necessary.

Artifacts, especially those from archaeological or ethnographic collections, are treated differently. This is because they may contain intrinsic information of value to future research, or because the maker's intentions may not be fully understood. Therefore, treatment is limited to the absolute minimum that will ensure the survival of the object.

This constraint becomes increasingly important as new research techniques are developed, because treatments that were thought to be correct in the past are now known to have either removed intrinsic information or to have so contaminated the material as to negate its value for research. Conservators were once taught to use "reversible" treatments; today we know that no treatment is fully reversible. While it may be possible to remove the visible evidence of treatment, either the materials used in the treatment or those used to remove them, will leave traces or will react with the original material in such a way as to change it permanently.[4]

Similar constraints apply to the conservation of natural science specimens. As biological species are threatened or extinguished, museum specimens, even of species that are numerous, become increasingly valuable for research. They are also important for exhibition, even though the two activities have mutually conflicting demands. A specimen suitably mounted for display may be little more than a chemically contaminated skin stretched over a plaster or fiberglass form, while an unprepared specimen stored for research may lose all semblance of its natural appearance. Even materials as seemingly inert as bones may react chemically with the museum atmosphere and grow crystals that cause disintegration; geological specimens may convert to other forms. For the conservator of natural science collections, the extent of acceptable restoration is a particularly difficult problem.

Museums of science and technology have a different dilemma. While they preserve artifacts, they also preserve the technology for which, and by which, those objects were created. Technology often has an abstract element that can only be expressed through demonstration. Therefore, museums are tempted to restore original objects to running order and to subject them to the wear and risk of use. The advent of highly successful "science centers," which unlike museums have no obligation to preserve original material, has encouraged a popular demand for the operation of artifacts. Although a very contentious issue for museums, it is a difficult one for them to resist. It is made more difficult by the mutual indifference that seems to exist (with some notable

Account book from 1609 with gilt leather binding damaged by light and heat. Det Kongelige Bibliotek, Copenhagen, Denmark.

Leather and Related Objects

Leather and other skin products are among the oldest and most versatile of man's materials. Durable in use but prone to decay, leather is especially vulnerable to microorganisms, atmospheric gases, strong light, and extremes and fluctuations of humidity.

Leather objects in museums, libraries, and historic houses require skilled and constant care.

Moreover, leather or parchment-bound books and leather furnishings may remain in use, despite their rarity. Inevitably, they suffer damage and require restoration.

Skin products may survive long burial in anaerobic (especially acid) conditions remarkably well. Such pieces, which are of immense historic value, demand special treatment techniques if their form and fine detail are to be recovered.

Petroglyph Canyon, China Lake Naval Weapons Center, Ridgecrest, California.

Rock Art

Man has engraved (petroglyphs) and painted (pictographs) on rock faces from the earliest times until at least the nineteenth century. Found throughout the world, these first graphic records of ancient man are today endangered by nature, but much more by man himself.

Sites have been damaged by vandals, obliterated by construction, and, more innocently, harmed by attempts to make them more visible or to preserve them with protective coatings. At some sites, whole sections of rock face have been stolen or removed to museums. At Lascaux, France, this most famous site of all has been closed because, after surviving intact for millennia, public visitation during the last century has disastrously altered the cave's environment.

Natural deterioration is insidious and less dramatic. Cycles of freezing and thawing, rain and drought cause rock to spall, while efflorescing salts and blowing sand erode the surface. A natural process—similar chemically to that of fresco painting—causes pictographs on limestone in wet climates to be concealed by an opaque layer of calcite.

The preservation of rock art is a complex challenge that requires both physical control and a special understanding of the ecology of the entire region. Attempts to treat the paintings alone rarely succeed and may well hasten their deterioration.

exceptions) between technological museums and the conservation profession generally.

The conservation profession must bear some of the responsibility for the situation in museums of science and technology and museums of natural sciences. They have remained largely outside our orbit, not perhaps because they wished to, but because we have failed to promote our cause. Today, some of them are anxiously seeking conservators, but there are very few professionally quali-fied practitioners who also have the technical skills that are required to serve such collections. Conservation training has neglected these fields, but training responds to demand. As a result, museums of science and technol-ogy frequently employ restorers trained only in the tech-niques of industrial maintenance, while museums of natural science rely on commercial taxidermists. In both instances, the staff rarely has any formal training in conservation.

3

The Role of Science in Conservation

The contribution of science to conservation has been pivotal. In bringing together materials research and the ancient craft of restoration, it precipitated the development of modern conservation. While this close association continues to be vital, it is sometimes uncomfortable. Scientists and conservators have different backgrounds, work to different criteria, and approach their common aim in different ways.

Science deals in measurement, produces quantifiable results, and enjoys the benefit of precision. Conservation applies those results to problems of infinite variety, the solutions of which have no absolutes. Every step depends upon the fallible judgment of the craftsman, and ultimately, upon his personal skill. In many ways, conservation is the antithesis of science. Einstein's remark (though made in a different context) is particularly apt: "Mathematics is all well and good, but nature keeps leading us around by the nose."

The applications of science to conservation fall into three broad categories: examination and analysis, deterioration and environmental studies, and research into improved methods and materials. Research in each of these areas responds to two needs: an immediate demand for information on work in hand (typified, perhaps, by a request for the identification of a material) and longer-term, though still "applied," research into a recognized problem (such as the treatment of waterlogged materials or the protection of stone monuments against acid rain).

León Cathedral, León, Spain.

Stone

In the manufacturing centers of Europe, stone buildings and monuments have been deteriorating since the beginning of the industrial revolution. On the great cathedrals, constantly shrouded in scaffolding, damage has been repaired by replacing disintegrating stone with new material. In the twentieth century, the problem has become more acute. Not only has the cost of replacement become excessive, but we have learned that a replacement is not a satisfactory substitute for a lost original.

The mechanisms of deterioration, both natural and manmade, are well understood. In the absence of chemical pollution, natural decay is normally slow, but air- and water-borne chemicals not only erode the stone surface, they also render it porous, thus further accelerating the natural process. With this knowledge, conservation efforts have turned to protection. If acid rain cannot be prevented, at least exposed stone can be given a measure of protection and, to some extent, stone already weakened can be strengthened. Although encouraging, such techniques are limited, piecemeal solutions applicable to only a fraction of the threatened heritage. As long as chemical pollution of the atmosphere continues, buildings, sculptures, and the natural environment will continue to erode.

A major achievement of conservation scientists in recent years has been the development of practical techniques, such as spot tests, which enable conservators to obtain for themselves information that formerly required an analyst. Simple ultraviolet and infrared examinations, straightforward radiography, and basic analyses are often undertaken by conservators, but the expertise of the scientist is the only resort for more complex problems.

Rigorous scientific examination of objects is often required for the purpose of authentication. Neither conservators nor conservation scientists will pronounce on the authenticity of an object, because such an opinion normally depends on historical or stylistic criteria, which are the domain of the curator. However, scientific information may provide the curator with the essential evidence for a conclusion. The evidence that the object contains may be sufficiently obvious to yield to the relatively simple examination of a conservator, but in more difficult cases, it demands the sophisticated techniques of the scientist. This is particularly so when the object is, or is thought to be, a skilled forgery.[5]

In environmental studies also, the conservator may require scientific support. Most conservators have a thorough understanding of environmental mechanisms which, together with the excellent monitoring instruments that are available, enable them to deal with most common environmental problems. Some issues, however, like the effective control of microclimates, often demand the attention of scientists specializing in that field.

Scientific research into improved conservation methods occurs in three phases. The first phase involves the identification of the problem, which may stem from an inadequate conservation treatment or a situation for which no suitable treatment exists. In the second phase, the scientist analyzes the problem and devises a theoretical solution. The third phase requires the translation of the theory into a practical technique and its subsequent testing by conservators, a process which requires close collaboration with scientists.

Long-term research is sometimes mistakenly regarded as a luxury, but it is, in fact, the fuel of future knowledge. We actually know very little about most objects in our collections. Existing identifications of materials in museum records are often based on nothing more than traditional assumptions that conservation examination frequently proves to be erroneous.[6]

Both the expertise and the equipment for conservation research are expensive, so there are as yet few facilities in which conservators and scientists can work together. This is particularly unfortunate because the close collaboration of conservators with conservation scientists in a common working environment has proven to be a most productive arrangement. Few museums can provide their conservators with in-house scientific support. Most must rely on the goodwill of the few major museums that have conservation scientists or on the assistance of industrial or university laboratories that have no intimate understanding of museum problems. Even in countries that have national conservation institutes, research services are invariably overloaded.

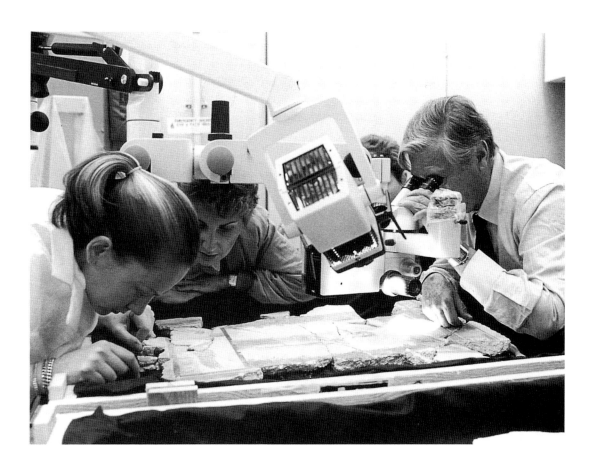

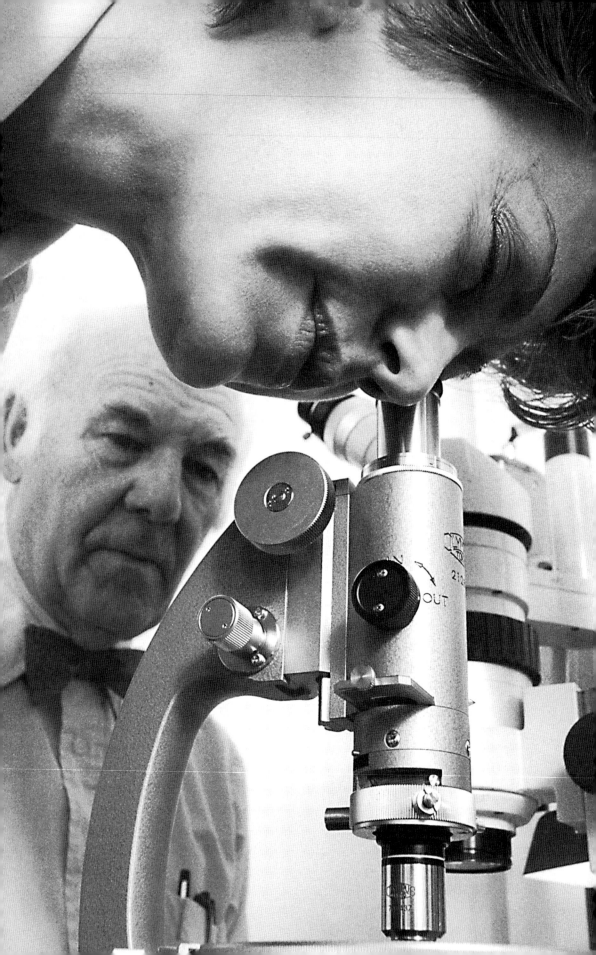

4

The Training of Conservators

It is in the field of conservation training that some of the most remarkable developments have occurred. Until the 1960s, almost all conservation training was acquired through apprenticeships. Museums trained their own conservators who often had an "arts and crafts" background. They received some training in basic science from the museum's own scientists or through courses at a college or university. The core of this training invariably consisted of intensive, practical working experience in the museum.

In some European countries there were long-established programs in restoration that evolved from the apprenticeships of the ancient craft guilds. These were generally specialized to the restoration needs of a specific craft and were often regional in application.

Another form of training was by apprenticeship to a private restorer. Apprentices worked in the master's studio and sometimes augmented their training with external academic courses. Training was always highly practical, though variable in quality. The methods (often regarded as trade secrets) that pupils learned were those of the master. The prestige of such training lay in the master's reputation.

In the 1960s and early 1970s, the changing philosophy of museums created a demand for conservators with more systematic training than these methods could provide. In particular, the need for a better scientific education and the wish of restorers to professionalize their craft led to the establishment of conservation programs at several universities in Europe and North America. Not only do these programs provide a sound background in environmental and materials science, but some universities have special programs for those seeking a career in conservation research. Previously, scientists came to conservation, as many still do, from either academic or industrial environments, with no more knowledge of conservation than the old restorers had of science.

There are also deficiencies in the training of curators. The work of conservators and curators is complementary; indeed, in some museological situations, it is almost symbiotic. Such collaboration depends upon mutual understanding at the professional level, yet very few of the academic programs from which museums recruit their curators offer even the most rudimentary training in the physical properties of the materials for which the future curator will be responsible. Consequently, the young curator rarely understands either how to care for his collection or the possibilities and limitations of conservation. There is an urgent need for those university faculties whose graduates may seek curatorial positions to offer basic courses in "conservation for curators" and for museums to require such training as a fundamental curatorial qualification.

A more recent painting over a sixteenth-century icon, being removed in one piece using a protective synthetic film, before its transport to a new support. VNIIR, Moscow, USSR.

Icons

Icons often present a special problem, not only because of successive restorations, but because fundamentally different images from succeeding periods have been superimposed over one another. The need for the detection, dating, and separation of images, without damage to any of them, has resulted in the development of special techniques. In particular, new ways to separate and manipulate the old, fragile paint films have been devised. This permits the analysis, cleaning, and restoration of both images and facilitates reproduction, using original materials and original methods.

On the other side of this issue, conservators and conservation scientists whose only work experience is in national or regional conservation service centers, may equally lack the museological experience to appreciate curatorial priorities. Ignorance of the other's discipline always carries the potential for misunderstanding between curator and conservator. It places a dangerous and unnecessary obstacle in the way of heritage preservation and it remains one of the unsolved problems of museological education. As a distinguished pioneer of conservation education wrote:

The fact remains that neither group knows enough about the other's expertise. The art scholar lacks in-depth knowledge of the nature and behavior of the material which has been fashioned into artefacts. The preservationist knows too little about the immaterial content of the artefacts he presumes to repair.

—Caroline Keck, "A Plea for a Practical Approach to an Old Problem," *Museum*, vol. 34, no. 4 (1982), p. 236.

Although the standard of training offered by university programs is high (usually leading to a Master of Arts degree or a postgraduate diploma) and fairly consistent, there are some philosophical variations between institutions. Some programs encourage early and intensive specialization, which produces conservators who are perhaps better equipped for sophisticated laboratories or private practice than for the more modest facilities of smaller institutions. Others offer a more broadly based training well suited to smaller museums, and some, like the Conservation School of the Royal Danish Academy of Art, encourage graduates to return to develop specializations after a period of work experience.

Many extremely valuable programs are offered at the second academic level of polytechnics, city colleges, and craft schools. They are generally, though not always, shorter than university programs, broader in scope, less theoretical, and less predictable in quality. While some museum administrations may view these graduates as technicians rather than professionals, many of them have achieved considerable success despite the lack of an advanced degree.

It is appropriate here to offer some definition of the difference in professional status between conservators and conservation technicians. This issue has proven troublesome in some administrations because it is frequently confused by bureaucratic attempts to classify employees on the basis of academic qualifications. The distinction ought to be based on experience, ability, and responsibility—not only on education. It is reasonable that the more highly qualified worker should have seniority, but in practice, academic qualifications do not necessarily reflect ability.

The conservator is a fully responsible professional, characterized by the ability to diagnose conservation problems and to design and carry out treatments without supervision. Although normally a specialist, he has a working knowledge of many fields and a broad knowledge of conservation theory and practice.

A conservation technician may assist the conservator by performing general treatments under supervision, or he may be able to perform limited tasks without supervision. He lacks the diagnostic skill and theoretical knowledge of the conservator, to whom he is responsible.

It is traditional apprenticeship that is the real casualty of modern conservation training. A few restorers still accept apprentices, and some museums and archives train technicians for their own use. But for the professional institutional conservator, apprenticeship is no longer a viable mode of entry. This is due in part to the increased emphasis on scientific literacy, which is difficult to satisfy in apprenticeship situations, and in part to its heavy demand on staff time.

It seems there is no ready solution. The increasingly technical nature of conservation demands a scientific background that can only be supplied by formal academic education; yet one must regret the passing of apprenticeships. Those who received such training acquired an understanding of museology and a commitment to its objectives that made them more comfortable colleagues than some of today's conservators.

There is, however, one way in which a form of apprenticeship survives. Many conservation training programs require students to undertake an internship in a reputable conservation laboratory as a condition for graduation. Such internships serve exactly the same purpose as the later years of the old apprenticeships: They provide intensive work experience in a busy institution under critical professional supervision. When the intern is well matched to the host laboratory the experience is mutually beneficial, and this exchange contributes to the realistic response of schools to the needs of future employers.

At a few major institutions, advanced internships are available to senior conservators to allow them to return to the laboratory after some years in the field to learn sophisticated new techniques. Both types of internships are generally arranged at no cost to the host institution which, in return, charges no fee. However, such opportunities are rare, and advanced training remains a problem.

Occasionally, it is possible to arrange mutual exchanges of personnel between institutions, even internationally. In general, the experience of those institutions that have participated in professional exchanges has been positive. Individuals, institutions, and the profession at large all benefit, and at modest cost. If economic and political conditions permit, this could become standard practice in the future.

The rapid technical developments in conservation demand that practitioners keep abreast of this evolution. For most, the only practical way to do this is by reading the journals of professional associations and by attending conferences. In the latter, conservators are handicapped by the cost of travel and the reluctance of many employers to support conference attendance.

Finally, there are the glaring holes in the system that have already been mentioned. The classical subjects—fine art, architecture, archaeology, ethnology, works on paper, and three-dimensional historical artifacts—are fairly well served by conservation training programs; furniture and textiles, rather less so. But the absence of qualified conservators specializing in natural science collections and in science and technology presents an acute problem to such museums, and one that is unlikely to be resolved within the existing training structure. Present conservation training programs owe their origins to the needs of art galleries and history museums and are usually located in university faculties of fine art. They may draw on existing courses in other faculties for such subjects as chemistry, biology, and business administration, but course work that specifically addresses natural science collections would require a fundamental reorientation.

Easel Paintings on Rigid Supports

Artists have painted on many materials other than paper, canvas, wood, and plaster. Sometimes this has been determined by the support itself, as when the motive was merely to decorate an object, but frequently it was because the artist sought a more permanent ground for his work. Ironically, such seemingly durable supports as metal, glass, stone, wood, and ivory have often proven to be poor grounds for the artist's chosen medium.

Movement of the support due to variations in temperature and humidity, attacks by insects and microorganisms, and chemical decomposition all threaten the adhesion of paint layers and render such paintings particularly fragile.

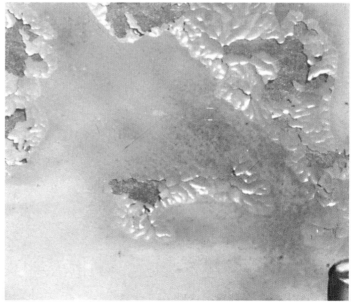

Eighteenth-century painting on glass, Bavarian school, private collection. Detail shows the detachment of the painting from the support.

Training for the conservation of scientific and technological collections presents an even greater problem. Such museums include not only general "science" museums, but also the countless specialized museums of industry, especially those concerned with transportation and communications. No training can embrace all fields of industrial technology, yet it is of paramount importance to such museums that their conservation staffs are expert in those techniques that are appropriate to their collections. To learn any one industrial skill to the standard required might take a craftsman a lifetime of experience at the bench. How then can a conservator do so? The answer may be that museums of science and technology must continue to employ craftsmen from industry who already possess such skills, but this work must be planned and supervised by a professional with formal training in conservation.

Banner with moth damage to wool portions. Nationalmuseet, Brede, Denmark.

Textiles

For all their delicacy, textiles often survive long burial, and even when only fragments remain intact, they are of great historical value. The identification of materials and the analysis of the weave may reveal information on trade routes, fashion, and weaving technology and may contribute to an accurate dating of the site.

In other cases, where a site is well documented, clothing found in graves—especially that of working people—provides much information about social conditions in general and about the lives of individuals in particular. Such pieces will rarely be restored, but they are usually cleaned, spread out or blocked, and mounted for storage or exhibition. They require care in excavation and transportation, and great skill in handling, washing, and mounting.

More recent textiles also require special care. Often they were used until they were worn out; then they were either crudely repaired, adapted to some other use, or simply thrown away. While numbers of christening robes, ball gowns, wedding dresses, and uniforms from the recent past survive, work clothes and other expendable textiles are extremely rare.

5

Conservation Services

Murals

Throughout Europe, from classical times, wall paintings were a feature of ecclesiastical and many secular buildings. The restoration of churches in northern Europe has led to the rediscovery of numerous pre-Reformation murals beneath the accumulated whitewash of centuries. In turn, the recovery and examination of such paintings has promoted further study of fresco techniques. The analysis of original fragments and experiments replicating the painting techniques and the various methods for preparing walls approach the same subject from different directions, but together improve our ability to preserve mural paintings.

Examination of the wall painting in Nefertari's tomb, West Thebes, Egypt.

The difficulty that some museums have experienced in determining the appropriate place in the organization for a conservation unit, often reflects a failure to understand the function of conservation. Fundamentally, the issue is one of centralization or decentralization. Should there be a single conservation department or should each curatorial department have its own unit? Museums developing their first conservation facility often choose the latter arrangement. It permits an economically modest start and it does not disturb the status quo. Curators may prefer it because it places control of the conservator in their hands and thus dissipates what some may perceive as a threat to their control of the collections. Conservation began this way in many older museums: not as institutional policy, but through ad hoc decisions by individual curators, which led to a multiplicity of small departmental units with different curatorial policies.

Whatever its origin, a decentralized system carries the potential for duplication, inconsistency, and, most seriously, weakness. With each conservator reporting to his curator, no single professional is responsible for an overall institutional conservation policy and, as a consequence, such museums often have none.

Centralization is infinitely preferable and, in the long run, more economical. It permits the museum to develop a unified policy, to standardize its procedures, and to avoid redundancy. The possibility of expensive and disruptive duplication is very real in a decentralized system. A large museum, for example, might find itself

Metals

Metal objects are always threatened by corrosion. When buried in archaeological sites, metals may be attacked by chemicals in the soil or by electrolytic reaction with other metals. Nor are they safe after excavation; minute traces of chemicals, especially chlorides, may react with atmospheric moisture to attack metal objects that have been stable for centuries.

Even metals that have never been buried may corrode after contact with organic materials, acid gases released by other materials, or water. Atmospheric pollution and high levels of relative humidity are always a danger. Microorganisms growing on adjacent materials may generate acids that will attack metals, and the fingerprints of handlers may be etched into polished surfaces by skin acids. Even modern display and storage cases in a controlled environment may contain paints and synthetic materials that release corrosive gases.

Metal corrosion can be prevented by controlling the environment, by the careful choice of display and storage materials, by the use of vapor phase inhibitors, and by vigilance. Even though corrosion can be arrested, the damage it causes can never be repaired.

Detail of first century B.C. Roman bronze bust with inlaid glass paste eyes, before and after cleaning with surgical scalpel. The J. Paul Getty Museum.

employing conservators specializing in ceramics, wood, stone, or metal in several departments. Conservation should be centralized, if only because it is normally the only materials-oriented department in the museum.

When the senior conservator is appointed to the same organizational level as the curators, he has the advantage of equality and may thus influence the practices of all departments in their handling, storing, exhibiting, and lending of collections. This is perhaps the most important benefit of centralization, and one which, when overlooked, most clearly demonstrates a museum's failure to understand the nature of conservation. There are still curators who think of conservators only as restorers (some, indeed, who think of them as artisans) and fail to realize that they are, in fact, the museum's only professionally trained specialists in preservation.[7]

In museums with fully integrated conservation departments, it is unusual for a conservator to spend more than half his time treating objects. The rest of his energy is devoted to environmental monitoring, pest control, object condition examination and reporting, assistance with exhibit design, technical examinations of objects for curatorial research, treatment records, and staff training. If the museum has an active field program, the conservator may be required to assist in the field. If he works for a regional museum supporting smaller institutions in its area, he may be expected to provide training and respond to emergencies.

Perhaps the most difficult situation arises when an established museum with a traditional structure appoints its first conservator. Curators may object to a conservator being appointed at their level. They may argue that their responsibility for the collections implies complete control, that restoration (which is what they often perceive conservation to be) is a manual, rather than intellectual activity (and therefore inferior), and that conservators do not have equivalent academic qualifications.

That these objections are spurious does not alter the fact that they arise from an ignorance of conservation for which conservators are themselves largely responsible. Conservators have always been eager to communicate with one another on technical matters, but they rarely explain themselves to their museological colleagues. They have failed almost completely to communicate the significance of their discipline, which in itself would justify their status in the museum. That significance is no less than one of the four fundamental responsibilities of the museum: *preservation*.

The conservator is most effectively employed when he is located in the museum where he can have daily contact with the collections. Initially, that was the way all conservators worked, but obviously, not all museums can employ conservators. With the growing recognition of the need for conservation on a worldwide scale, some systems have been developed to provide conservation services to museums and other institutions that cannot support a conservator on staff.

Through simple ad hoc arrangements where large institutions helped small ones, more formal systems evolved. Sometimes these were forerunners of integrated national conservation structures. In their most sophisticated form, such structures include all the elements of a mature profession: treatment services, scientific support, training, professional accreditation and registration, advisory services, an information network, professional associations, and emergency services.

These systems generally reflect the political character of their country. They vary from loose, informal associations of autonomous organizations linked only by common purpose, to those linked by a benign national policy and governmental project funding, to others that are wholly government controlled.

In developing countries, the absence of a resolved system may be due less to poverty or strife than to the absence of a "critical mass." In some developed countries where a substantial conservation structure has evolved only recently, growth was very slow until a sufficient level of activity began to generate its own momentum. When that point was passed, and the political and economic climates were favorable, the remaining parts of the structure tended to propagate themselves.

No two national structures are alike. Even those that are comparable in terms of development may be fundamentally different in process. Some are inclusive, embracing not only museums, art galleries, and archives, but also historic buildings, historic and archaeological sites, and sometimes the natural heritage as well. Others, perhaps for historical reasons, separate archives from museums, and museums from heritage buildings; or they associate art galleries with archaeology, or museums of natural science with agriculture or mining. In any event, whatever the organizational nature of the structure, if it evolves, it is likely to undertake many of the same activities.

Large countries face the challenge of providing conservation services to communities separated by great distances. In the United States and elsewhere, this has been achieved by establishing regional conservation centers, while in Canada mobile laboratories have proven more successful.[8]

Another service that conservators have long provided to institutions other than their own, is training in the care of collections. Through seminars often organized by national or regional museum associations, care of collections training covers all aspects of artifact care short of actual treatment. This is perhaps the service most valued by workers in small museums.

Closely related to training in the care of collections is the provision of information on the same subject. A substantial and growing literature exists, written by conservators, specifically for those who handle artifacts.[9] Such works range from books to paperback "technical bulletins," to one- or two-page information sheets. Some national conservation institutes distribute these publications free to all museums in their jurisdictions.

A national conservation structure cannot achieve maturity without scientific support to address the special problems of its region. Such support may be provided by universities, by the laboratories of a national museum, or by a national conservation institute. The essential features are that it be innovative, accessible, affordable and of high quality. The best scientific support services have the ability to balance their response to the immediate needs of conservators with a creative and unconventional approach to the anticipated problems of the future.

A common difficulty of conservation systems in the early stages of development is the shortage of conservators, who must often be recruited abroad. This practice can only be a temporary measure; sooner or later, a nation must train its own conservators. This will depend on the response of the universities because the scope of conservation is now such that it can only be taught effectively where courses are already available in a wide range of disciplines.

Professional associations such as the ICOM Conservation Committee and the International Institute for Conservation of Historic and Artistic Works (IIC) ensure

Two identical glass drinking cups, c. 1700; effects of crizzling can be seen on right. De Danske Kongers Kronologiske Samlinger, Rosenborg Slot Copenhagen, Denmark.

Glass

Glass is not a true solid. For all its hardness, it is really an *extremely viscous liquid—sensitive to pressure, temperature, and atmospheric moisture. Correctly formulated, glass is extremely durable, but some ancient glasses are unstable and respond to high relative humidity (RH) by "weeping" or to low relative humidity* by "crizzling." As it is rarely practicable to correct these conditions by treating the objects themselves, they are controlled by providing special environments for the affected pieces.

that even in the most isolated locations, a conservator is able to keep in touch with world developments in conservation and with colleagues in other countries. When sufficient conservators are present it is likely that they will form a national group of the IIC or a specialized group within the ICOM National Committee. This is a vital step in the development of a national conservation structure. The most important function of conservation associations is the exchange and dissemination of technical information, which benefits all members by ensuring their access to quality information and encouraging uniformity of practice.

One feature of a national conservation structure that is often late in developing is an emergency service. Museums affected by major emergencies can usually be assured of assistance from other museums, but that is a poor substitute for an organized emergency response by trained professionals. There is one point in any major cultural emergency at which on-site conservation expertise is crucial: when recovery of the collections begins. If the collections are not to suffer unnecessary additional damage, it is essential that their recovery be directed by a conservator. Approximately ninety percent of museum emergencies involve water damage, which demands immediate, skilled care if objects are to be stabilized successfully. The concern at this stage is not restoration but stabilization, so that objects do not deteriorate any further before they can be restored. This is where the emergency response team is so important. Conservation facilities should plan with museums in their regions to make small teams of conservators available to respond quickly to any kind of cultural emergency.[10]

6

The Conservator as Museologist

Conservation is the latest major discipline to enter the broader field of museology, and most of its difficulties stem from this fact. Conservation is still developing and, like any newly introduced species, conservators are still discovering new niches to occupy. Because it is concerned with all collections and strays into every museum activity in which collections are used, conservation has not always been welcomed or understood.

Museums were sometimes slow to recognize the potential benefits of the new discipline. Exhibit design was often established as a force in the museum long before the first conservator was appointed. Exhibits were planned and built without conservation advice and if the newly arrived conservator attempted to point out errors, he would be advised to not interfere. If he persisted, he would be told to wait until the exhibit was finished before commenting. By the time he was heard, the exhibit would be a fait accompli and already popular with the public. Any criticism made then would be dismissed as unrealistic because the faults would then be too expensive to correct. Ergo, the conservator is typecast as an interfering nuisance with a negative attitude.

Fortunately, this rarely happens today, but many older conservators can testify to its truth. In progressive museums, the conservator is part of the exhibit design team. Thus, by defining the parameters for a safe exhibit environment and by helping the display staff achieve them, he can ensure that the exhibit is environmentally acceptable without expensive alterations midcourse. Good designers welcome such assistance and find no difficulty with environmental criteria that are set at the start. It is rarely more difficult or more expensive to design to good criteria than to poor ones. In addition to establishing criteria for the welfare of objects in the exhibit, the conservator must be willing to assist in the solution of technical problems by experiment and testing. This is a particularly interesting aspect of conservation because as designers constantly try new exhibit methods and materials, new challenges arise with almost every display.

There has always been a bond between conservators and registrars, who share a common concern and responsibility for the physical welfare of collections. Often it has been the registrar, faced with the problem of organizing and storing diverse objects (usually on a small budget and under crowded conditions in the least suitable parts of the building), who has first recognized the assistance that a conservator can provide. The conservator's help in diagnosing the causes of deterioration in storage and finding solutions for them; in improving the storage environment; in designing special supports and containers; in implementing effective pest control; and in handling, packaging, and the transportation of objects is welcomed by registrars.

On the other hand, museum educators may be wary of conservators. The two disciplines tend to view collections from different standpoints: one sees them as tools to be used, the other as resources to be preserved. The two points of view are not irreconcilable, but solutions demand compromise and understanding. Furthermore, the answer, whether it is to form a separate teaching collection, to build replicas, or to use audiovisual techniques, is likely to be expensive and therefore unwelcome.

Although the difficulties that occasionally arise between conservators and curators have been mentioned several times, they are by no means the norm and it is hoped that these comments will be accepted in the constructive spirit in which they are intended (The author has been a curator himself!). When problems do occur, they are usually due, in fact, to the very close philosophical and operational relationship between the two functions. At its most fundamental level, the relationship is one of shared responsibility: The curator is responsible for the intellectual aspect of the collection—for what it

Ethnographic Materials

Ethnographic collections encompass perhaps the widest range of materials in museums, potentially including any material that has been worked by man. Consequently, the range of problems that may confront the conservator of ethnographic collections is correspondingly vast.

The commonplace problems of all collections must be addressed: the deterioration of objects through poor storage and handling, the effects of inadequate environmental control, dirt, and insects. However, more perhaps than any other kinds of collections, ethnographic materials demand that special attention be given to the design of suitable cases and supports for storage and exhibition. Objects that are of exotic form and fragile material, decorated with fugitive pigments, and intended to be expendable (but are nonetheless irreplaceable), place the ethnographic conservator in a special relationship with the exhibit designer. Only a close working collaboration between the two can avoid unintentional, but serious damage.

Mambilla figure, wood, Nigeria/Cameroon. UCLA Museum of Cultural History, Los Angeles.

Facsimile of the Lascaux Cave. Wood supports were generated from photogrammetric analysis. Photo-transfers of the original paintings were applied to the resin-covered supports.

Facsimile and Conservation

Some museums replicate objects in their collections for purely commercial reasons, but the many complex techniques of facsimile production also serve the purpose of preservation.

Archives routinely copy documents onto more durable materials, not only because the original medium is deteriorating, but also to provide expendable copies that may by freely used for study or display. Reproductions may be made for educational purposes, so that those who cannot travel to see the originals may at least see replicas, or so that students may handle facsimiles of originals too fragile for such use. Exceptionally, whole endangered sites, such as the caves at Lascaux, may be reproduced elsewhere so that access to the original site can be restricted.

More often, damaged objects that are too fragile to be restored are replicated. A replica can be restored to show the intended form of the original, or parts missing from an original piece may be restored with replicas copied from similar parts of the same or another object. Copies are sometimes made so that alternative methods of restoration may be tested or demonstrated. In electron microscopy, certain subjects can only be examined in replica form because the electron beam would destroy original specimens.

Although replication techniques are becoming increasingly sophisticated, they must be used with care. Some widely used methods frequently cause damage to the originals that is not always immediately visible. There are also ethical considerations: If replicas are good enough to represent the originals accurately, yet are freely circulated because they are only replicas, might they not be misused? Should the museum, whose existence is founded on truth, allow such deception, even with the best intentions?

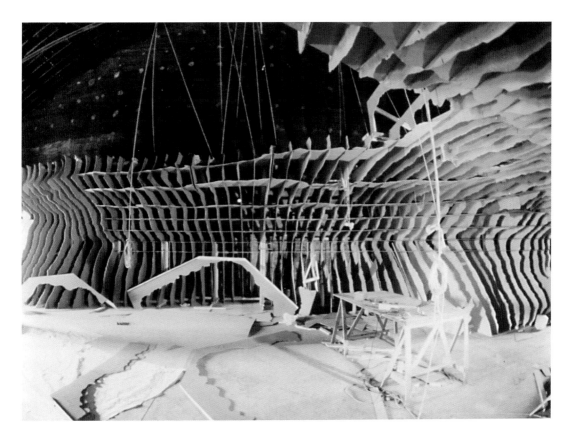

means—while the conservator is responsible for its physical aspect.

Friction sometimes occurs because, as we have noted, conservation is very new and some curators, sole masters of their collections for so long, view the conservator's interest as an intrusion, or even a threat. It most assuredly is neither. The conservator does not challenge the curator's custody of the collection and questions his use of it only when it threatens the collection's welfare. That, surely, should be a cause for satisfaction, rather than concern.

On the other hand, conservators cannot deny their share of the responsibility for misunderstandings. We have not always appreciated curatorial priorities and sometimes have pressed our case beyond reason. The conservator's excessive zeal is due, perhaps, to his need to establish his presence in an institution that had always believed it was doing very well without him. While this may be a valid reason, it is hardly an excuse.

In most museums, happily, conservation has become accepted as a mature and essential discipline that complements the work of curators. In some institutions it is standard practice for all proposed acquisitions to be examined by the conservator before acceptance. His technical examination may reveal an inherent vice that was not apparent, or may find some anomaly in fabrication, material, or evidence of use that casts doubt on its provenance or authenticity. Similarly, his skill in interpreting the physical features of objects in the collection may be of great assistance to curatorial research. Such examinations have frequently resulted in the confirmation of doubtful attributions or the revision of established ones.

Collaborations such as these are helpful and satisfying to both conservators and curators. They invariably strengthen the museum professionally and, interestingly, they lead to a kind of cross-pollination between the two disciplines that demonstrates their natural compatibility.

In other ways too, the conservator may function as a general museologist, beyond the range that many expect. Because of the toxic, flammable, or corrosive chemicals they use, conservators receive considerable training in laboratory safety, a fact which not only enables them to advise other museum workers (who are rarely as well informed, even though they may use materials just as hazardous), but also often leads them to take an active role in museum safety activities.

Again, it is often the conservator who, perhaps in concert with the registrar, takes the initiative in emergency planning. The reason for this is obvious. Minor emergencies, such as water leaks, are commonplace and, inevitably, it is the conservator who has to repair the damaged objects. Most accidents of this type can be avoided or at least minimized by adequate planning. Many a museum's emergency procedures have evolved from nothing more complex than the conservator leaving his home telephone number with instructions to call him if an accident occurs.

Thus, the conservator, simply by the nature of his craft, often fulfills the role of a museological generalist. To quote Jean des Gagniers again:

Il reste que, quelle que soit sa spécialité, un professionnel de la conservation possède des connaissances générales qui dépassent de loin celles des autres muséologues.

It remains that, whatever his specialty, a professional conservator possesses general knowledge that goes far beyond that of other museologists.
—Jean des Gagniers, La Conservation du Patrimoine Muséologique du Québec (Québec, 1982), p. 11.

7

The Future

Although not itself a science, conservation recognizes science as one of its parents, to whose destiny its own is inescapably linked. The frantic pace of technical development in conservation is the result of that union and it presumably will continue for as long as the impetus of scientific discovery is maintained.

In some respects, the one not only forces the other, but also points its way. For example, the challenge of synthetic materials in our historic collections will certainly be a major preoccupation of conservators and conservation scientists for the foreseeable future.

Industrial plastics have been used in Western industry since the nineteenth century, but the formulation of those early materials was relatively simple.[11] Since the middle of the twentieth century, however, synthetic materials of great complexity have been in widespread and increasing use in countless domestic articles that are now finding their ways into museum collections. The formulations of such materials are changed frequently and without notice and, moreover, they are often designed to have a limited life. Thus, we must contend with not only the natural deterioration of materials in response to the environment, but also with the "planned obsolescence" manufacturers deliberately build into their products.

Obviously this poses an immense technical challenge for the conservator and thus for the scientist who is required to analyze materials and devise ways to arrest their deterioration. But it also presents an ethical dilemma that sorely tests our wisdom. It is one that is already familiar to conservators who have confronted the

Photomicrograph of scratch on glass using cross-polarized light and 1/4-wavelength filter.

problem of auto-destructive art, but since it encompasses entire categories of historical objects, it is infinitely more serious. Put bluntly, if eventual deterioration was the intention of the maker, is not that deterioration a part of the object's integrity? Can the conservator legitimately interfere with it?

If this present narrative has a theme, it is that conservators must learn to communicate with other museologists. The urgent need for the members of this young, highly technical profession to share information has led them to adopt the terminology of science, which is singularly unsuited to the humanities. As Madeleine Hours has pointed out, the conservator is at the very center of C.P. Snow's well-known dilemma:

We must avoid the separation between a scientific culture, only too often content with indisputable proofs which we now know can be challenged, and the humanistic culture. . . . The establishment of a dialogue implies respect for others' ways of thinking; it will therefore also be necessary to make a moral effort, and to rethink our vocabulary and language.

—"Origins and Prospects,"
Museum, vol. 34, no. 4 (1982),
p. 23.

Until we rethink our vocabulary and language, the closer association with other museologists, which we all desire, may continue to elude us.

Communication in a different form will figure largely in conservation in the future. The Conservation Information Network currently being developed through the collaboration of the Getty Conservation Institute, the Canadian Conservation Institute, the Canadian Heritage Information Network, the International Centre for the Study of the Preservation and the Restoration of Cultural Property (ICCROM), the International Council of Monuments and Sites (ICOMOS), and the Smithsonian Institution will certainly lead to a more extensive linkage of computer-accessible information resources. Eventually we may expect a worldwide network, through which all major conservation facilities may use and contribute to a common pool of information including, but not restricted to, conservation literature and materials used in conservation practice.

Documentation

Information is the fuel of conservation. The conservator needs to know as much as possible about the materials and fabrication of the object he is treating, and about all the previous treatments it may have received. The curator and the historical researcher must also know exactly how the conservator's treatment may have altered the object.

In the course of treatment, the conservator will require information about many materials and their behavior: not only the materials of which the object is made, but also those he may use in treating it. He should have the best possible access to the vast amount of published material in any language and also to the results of

relevant research that may not be yet published. Hitherto, this has been physically impossible, even for those in the largest centers with extensive libraries and sophisticated equipment. Now the computer brings access to all this information within our reach. The Conservation Information Network (CIN) now being developed by the Getty Conservation Institute, the Canadian Conservation Institute, and the Canadian Heritage Information Network has the potential to provide such a multilateral information network of users and contributors.

Gaps in our training system will be filled because they must be filled. It is simply no longer possible to ignore the conservation needs of museums of natural sciences and museums of science and technology. Neither is it acceptable to have curators who have no knowledge of the physical needs of their collections nor conservators with no understanding of curatorial priorities.

Conservation will necessarily evolve into different forms. It is possible that the two major activities, prevention and restoration, may become separated in some institutions. As the preventive function of the conservator assumes greater importance, some museums may choose to employ him as a full-time preservation specialist, rather than one who functions as a part-time restorer. This is unlikely to become the norm, but some museums may prefer to eliminate expensive restoration facilities and contract their restoration needs to private sector restorers, working under the direction of the institution's conservator (preservation specialist). This is not entirely speculative, for many senior museum conservators already specialize in preventive conservation, if only because their duties no longer allow them the time for major restoration projects.

Clearly, such a role for the conservator would require the closest collaboration with curators and indeed, an intimate understanding of their disciplines. It represents the mirror image of the situation proposed above, in which the curator of the future will have a working knowledge of conservation. It is one of the ways in which the two professions may move closer together.

These are only a few of the ways in which conservation may evolve. It has already come far since that day, only fifty-five years ago, when the word was first applied to cultural preservation. Today, it is a worldwide profession: vigorous, technically diverse, but philosophically consistent and devoted to the single aim of preserving the cultural heritage. The immediate motive for preservation is irrelevant: study or display, education or entertainment are only purposes of the moment. What matters is that the objects they use should survive.

Why is this survival so important? Why preserve the things that have passed? Do we care what Tyrannosaurus looked like? Do paintings and sculpture touch our lives? Does it really matter that our children hear the creak of a wooden ship?

It does matter because these are the memories of our human progress. The future is a void, and the present, a fleeting reality that slips instantly into the past. Our heritage is all that we know of ourselves: what we preserve of it, our only record. That record is our beacon in the darkness of time; the light that guides our steps. Conservation is the means by which we preserve it. Like the museum itself, it is a commitment not to the past, but to the future. "The rear-view mirror is our only crystal ball—there is no guide to the future except the analogues of the past" (Northrop Frye).

Notes

1. Since this book was originally written in English, it may be
   helpful to offer an explanation of the conservation terminology used in English-speaking countries, and the ways in
   which it differs from that used in Spanish- and French-
   speaking countries.

   In general English usage, "conservation" is synonymous
   with preservation and, as used in the museum field, it
   embraces both the prevention of deterioration and restoration.
   Similarly, a "conservator" is a professional practitioner in both
   fields, while a "restorer" specializes in restoration. Because the
   distinction is less clear in other languages, some international
   bodies use the term "conservator/restorer" to embrace both
   activities. The term "conservationist," often wrongly used by
   journalists, means "one who supports or favors conservation,"
   not one who practices it.

   In French, "*la conservation*" has a meaning similar to the
   English "preservation" and "*un professionel de la conservation*" is
   a conservator. "*La restauration*" means "restoration" and
   "conservation" in English, and "*le restaurateur*" is applied to
   both conservators and restorers, because "*le conservateur*"
   means not a conservator, but a curator.

   In Spanish, "*conservación*" refers to both conservation and
   preservation. Conservators and restorers are described by the
   term "*restaurador*," while a curator is called "*conservador*."

   There is no exact French or Spanish equivalent for the
   English "registrar," but many Canadian institutions use the
   rather outdated term, "*registraire*." In any event, the registrar's
   responsibility—documentation—is the same in both
   languages.
2. See Philip Ward, "In Support of Difficult Shapes," *Museum
   Methods Manual 6* (1978), p. 2. Published by the British
   Columbia Provincial Museum, Victoria.
3. The new technique of leaf-casting, however, allows the exact
   replication of areas of lost paper. This is particularly valuable

because it allows badly damaged books to be rebound and separate sheets to be stored safely.

4. It was once common practice to wash archaeological finds on the principle, then widely accepted by museums, that "patina is important, but dirt has no historical value." Recent research has shown that stone weapons, even after burial for thousands of years, may retain on their surfaces sufficient hemoglobin for modern analytical techniques to determine the species of animal killed or butchered with them.

5. Such examinations are not conducted exclusively on behalf of museums. In Canada, for example, the Royal Canadian Mounted Police, whose forensic laboratories are highly sophisticated, often refer art fraud cases to the Canadian Conservation Institute.

6. The precise identification of the materials of large numbers of North American native artifacts, now being undertaken by the Canadian Conservation Institute, has already shown some previous assumptions to be false. Not only will the resulting database provide essential information for conservators treating such objects, but it will also prove of immense value for curatorial research.

7. It is significant that museums which have only recently employed their first conservator frequently express surprise and satisfaction with the influence the new employee has on everything from storage and handling procedures to exhibit design and workshop safety.

8. The Canadian Conservation Institute uses light trucks consisting of a 4 x 2 chassis carrying a custom-built body: 4.27 m long, 2.4 m wide, and 2.13 m high. They are climate-controlled and come equipped with work benches, fume extractors, sinks, and storage for the equipment and materials necessary to carry out basic conservation treatments. They usually carry a staff of two: a conservator from the Institute's staff and an intern.

   After a successful pilot project in the Atlantic Provinces in 1979, five such vehicles have served Canadian museums from the Atlantic to the Pacific and from the U.S. border to the Arctic for seven years. Their impact upon the conservation awareness of the museums they visited has been remarkable. Beyond the limited treatments that can be undertaken with such facilities, they have given many museum employees an understanding of simple conservation measures that now enable them to help themselves. This program has so improved the capabilities of small museums that in 1987 it will be replaced by a program of team visits better suited to the needs of museums that are now distinctly more sophisticated.

9. The subtitle of one of the first (and still one of the best) of these books defines the whole genre: *The Care of Historical Collections: A Conservation Handbook for the Nonspecialist* by Per E. Guldbeck, American Association for State and Local History, 1974 (rev. 2nd ed., *The Care of Antiques* and *Historical Collections*, 1985).

10. The Canadian Conservation Institute, which has provided such a service for many years, can place a small team with the necessary equipment, on-site, anywhere in Canada, within twenty-four hours. Recently, in the United States, the Getty Conservation Institute has initiated studies on natural disaster planning, mitigation, and response on an international level.
11. Some would date the industrial use of plastics from the Chinese discovery of lacquer in the first millennium B.C.

Editing and coordination: Irina Averkieff
Design: Joe Molloy, Santa Monica, California
Typesetting: American Typesetting, Inc., Reseda, California
Color separations: Color Service, Inc., Monterey Park, California
Second printing, 1989: Westland Graphics, Burbank, California

Photography: Guillermo Aldana 8, 12, 44; Steen Bjarnhof 41; Stephenie Blakemore 15; C. William Clewlow 26; N. Elswig 50; Dieter Goldkuhle 30; Jesper S. Johnsen 21; Dingeman Kalis 17, 61, 63; Dick Meier 22; Thomas Moon 33, 46, 52; Berit Muller 43; Jaime Nuñez 10, 11; Knud Rasmussen, 5th Thule Expedition, Greenland 18; Knud Søndergaard 5; Paul Slaughter frontispiece, 28; Richard Todd 55.